E PLURIBUS UNUM

"21st Century Homestead"

How Aquaponics and CNC Free

Hydrogen Rodeo LLC

Version3 published from Texas, 2022

Buck@H2Rodeo.com

Acknowledgments

To my family, who have endured years of SenatorMark4 jibes from all directions. May they continue to be quick on their feet!

A driving force to release this new version of the book was the inspiration that Elon Musk and SpaceX have provided to reinvigorate the idea of space exploration and settlement. The final chapter about mining the asteroids was a non-trivial dream when I first wrote this, but now that space launches are becoming routine and we've actually returned asteroid dust to the planet surface (Japan's Hyabusa-2, December 2020) things have changed. Now we're talking engineering problems of proven systems.

You and I could live to see the rain of material wealth landing from asteroid 16 Psyche! It was discovered on 17 March 1852 by the Italian astronomer Annibale de Gasparis from Naples and named after the Greek mythological figure

Psyche. It is a very real possibility that on the 200th birthday of its discovery we will be mining it efficiently, sparking the new Golden Age, an age of infinite possibility.

Thank you, Elon!

My most heartfelt thanks go to my Navy buddies Pat and Amy. They inspired me to reissue this.

Contents

Designer...Thanks!

Bio

Carol Oquinn-Boyd

My love of art, design, and the creative process began at a very young age but was not realized until I was well into adulthood after completing a course in computer graphic design. I was invited to start a company with a partner as a web design company, my half not being the coding part. We gained marginal success during our first six months with a handful of clients when he decided to pursue other interests. Herding cats!

I couldn't walk out on our obligations to the clients we had so I learned HTML when it was the new thing and there was no support.

I still work in the field fifteen years later, however it is the designing side of it that I really gravitate towards. I have done work for magazines, newspapers, and various corporate clients.

The fine arts are another passion. I love painting, drawing, pastels, acrylics, sidewalk chalk...you name it. I do portraits and murals on commission and sell my work online and at Trade Days in Fredricksburg, Texas.

For more information please visit my blog at CarolOquinnBoyd.blogspot.com.

Preface

The world is such an exciting place today. I reworked this version because in the six years since it's original release things have really changed. Previously, I spent lots of time giving examples of the power of Computer Numerical Control (CNC) and 3D printing so people could start to believe. Now people are routinely making parts for antique cars, printing actual homes, and even experimenting with printing ribeye steaks. What a world!

The fact of the matter is that there are thousands of things happening now that will bend the future of mankind towards something that nobody can predict. The goal of this little pamphlet is to enhance awareness of opportunities for mankind to learn to live more like family.

In this time where all societies are so heavily impacted by the spread of the Chinese Coronavirus, the ability to be a self-contained family unit has never been more important. If the government is going to force you to stay home, not allow you to work, and you don't have the ability to feed your family...???

We can care for each other like neighbors again, helping harvest the bounty of the universe.

A Jeffersonian Ideal

In **"Notes on the State of Virginia"**, Thomas Jefferson stated clearly that he believed that "those who labor in the earth are the chosen people of God" and because they depend on their "own soil and industry," are the repository for "substantial and genuine virtue". He also suggested that factory workers, since they were dependent for their job through their employer, led to "subservience and venality" and that those types of jobs destroyed "the germ of virtue". To say that he felt that city living and the dependence it breeds as regards your requirements for daily living, which reside outside your control, was bad is simply an understatement. Jefferson makes no bones about the thought that this type of living deprived the citizenry of the type of independence they needed to develop the character traits that were absolutely required to infuse a nation's virtue. In QUERY XIX, "The present state of manufactures, commerce, interior and exterior trade", he says that **"the mobs**

of great cities" do for the strength of a nation "as do sores on the human body".

Like most people my age, our parents were raised up during the Great Depression and yet they managed to raise huge families. Perhaps there weren't always new shoes, maybe there was a shortage of funds for things like candy, but the family generally stayed fed. Shooting squirrels with a slingshot, using only one .22 caliber bullet per day because of the expense, and creative trapping managed to keep the beans and cornbread as side dishes for most meals. And yet, even beans and cornbread, strengthened by the eggs from the barnyard, is a real meal. A growing family was proof of love and wealth from heaven. Great Depression? Who noticed when your finances before or after were indistinguishable?

It's easy to understand in today's more urbanized world that we are all more fearful about being able to raise our families. The tendency to work the extra days and hours in order to secure your place in an organization which is being battered from abroad does not make for a joyous citizen. A contented parent needs to share a peaceful moment with their children and spouse. And, of course, if you filled the pot with squirrel using a pellet gun PETA would be knocking your door down with armed men!

Today, speech is infused with the phrase "social justice". Most of these programs are nothing more than big government mechanisms talking about trying to guarantee a fish in every pot. While this type of electoral positioning is sure to gain some traction with a large portion of any electorate, it is not a position that has any long term future. If you can't count on yourself, you are under somebody's thumb.

My view is that the screaming masses in any city are the result of the failure of the imagination to focus on the actual technologies that we have available today. Unlike Francis Fox-Pivens, the activist who basically called for riots in the streets to force governments to hand out fish to people, I believe we've reached a cusp event in history that eliminates the need for that kind of social agitator. Like most people from her side of the political aisle, the government can never do enough so it must be collapsed to enable even more government. This is not the American way.

The internet and aquaponics are the two most important tools that average people need to secure a future. Instead of advocating violence and promising 'income,redistributed' to voters, isn't it really the truth that social justice exists when you can feed and educate yourself? When you can feed yourself you don't join riots for bread. When you can educate yourself there is no limit to your future potential!

If we are truly empowered by a Constitution that guarantees you the right to life, liberty, and the pursuit of happiness and through your own efforts you can feed and educate your family...Isn't that the dream we can all embrace?

He saith unto him the third time, Simon, son of Jonas, lovest thou me? Peter was grieved because he said unto him the third time, Lovest thou me? And he said unto him, Lord, thou knowest all things; thou knowest that I love thee. **Jesus saith unto him, Feed my sheep.**

John 21:17

Introduction

Rapidly evolving technology seems able to bring so many things into our lives that mere months ago appeared to be dreams from a comic book artist. Yet with all this exploding material wealth, the desire to have a family and share the love of life is being oppressed. A home mortgage gives only a roof and lifelong promise to labor, if possible, to pay it off. The garage is but a void representing the requirement for transportation payments to get to the places that supply us with the money to purchase our living staples. In many respects, we really are living in prisons of our own making.

The goal of the "21st Century Homestead" is to allow people to be self-sufficient. The ability to harness technology to concentrate our lives into our home, our personal castle, and make us more independent of the forces outside our family is the ultimate goal. Allowing us to free our imaginations and let each castle be as unique as its owner is a compounding benefit. The

Jeffersonian ideal of the self-sufficient citizen being able to understand and support self-governance can and should be undertaken. It is possible in this ever expanding, spiraling upward, technological sweep and it must enshrine supporting humanity as its sole purpose and driving force.

This short pamphlet is designed to appeal to those who want our society to be a more people-friendly design. Architects and engineers are sure to find plenty they would change or note errors. A vision always has room for change because it is nothing but smoke in the beginning. This smoke is trying to ignite the fire of innovation for our pursuit of happiness!

Designers and decorators should recognize a place where their vision is unlimited using current technology. These are the people that mold our internal living spaces that gives us joy when we open our door. We aren't even at the beginning of the kind of beauty and creativity that will be available to us in the very near future.

People looking for a place to raise their families can hopefully see the germ of a vision for a happier place in these words. We are at a time where each of our contributions can be magnified immediately to everyone's benefit. If squirrel and dumplings is no longer an option and has been replaced by 3D printed ribeye steak for everyone, we're moving in the right direction. We can all push without trying to collapse anything we've already built.

Future Housing

"It is not the strongest of the species that survive, nor the most intelligent, but the one most responsive to change."

-- Charles R. Darwin

This is one of the most exciting times for anyone to live. My grandmother, born in 1900 in West Texas, was always talking about the changes she could remember before she passed away in 1992. She could recollect the excitement as men first started to fly as well as the first dusty footsteps on the moon. My grandfather on the other side of the family was four years older than her and was a blacksmith from a family of blacksmiths. The thing I remember about his impression of technology was that he would mumble under his breath over and over again, in apparent complete amazement, "A Mile A Minute", whenever we were cruising down the highway. Can you imagine?

Those advances are nothing compared to what we are going to encounter in our travels through life. But even with this great technological advancement washing over us, there will still be laundry. Our homes. The meals with family and friends. This is the area of life that has never changed, but our ability to enjoy these small pleasures is what has suffered greatly over the last few decades.

It is getting harder and harder to not believe something very significant is happening in society as relates to the housing of our families. The failure of government-driven solutions for housing has dented our enthusiasm for a home mortgage that owns us in times that have to be called "interesting". Public housing and it's hive of caves shouldn't even be in any discussion about "housing" for human beings. The job market is making the acquisition of sufficient food dependent on systems easily disrupted by natural and man-made disasters. There has to be a better way to make a life for ourselves and our children without bankrupting the country and living in fear for our daily bread.

I don't understand why we haven't advanced quicker since the 1968 Hemisfair, when the Palacio Del Rio hotel in San Antonio, Texas went up with 500 rooms in 202 working days. That was a time before the Apple computer. It is past time for man to build with the speed of our current gigabit technology and service the needs of our crowds with automated urban building, and farming, solutions.

One of the many really exciting technologies that is in full bloom now is the field of 3D printers. In the simplest example, it is nothing more than the technology your Mother used to create that pumpkin-shaped cake for your Thanksgiving. By building up layer upon unique layer, the eventual shape of the pumpkin was developed. Any wedding cake exposes some of the heights that this approach can attain.

With 3D printing it is exactly the same. The only difference is that the layers are much thinner but it can be many, many kinds of materials just like a cake can be chocolate or cherry. We are at the stage of 3D printing technology where limits are being tested by imagination. Printing houses, printing living organs for transplant, and as mentioned earlier (and dear to my heart) printing ribeye steaks is something that is happening now. There are 3D printing solutions using thermoplastic that are cheaper than the first IBM PCs when they first hit the market. We all watched that growth in astonishment.

The previous examples were really just trying to pique your interest. The charge of this 3D printing technology has already permeated so many fields that it is really impossible to list them all. Google "3D Printed House" and stand back.

Building a house with unique, robotically-engineered parts is now easy to see but it is much more interesting to imagine what the interior spaces are going to look like when a homeowner can literally print out their interiors. Design it yourself or download the patterns from the internet! Any style from any epoch, is now ours to enjoy. We are on the cusp of having an unlimited supply of raw material to use.

CNC your Vision

While the technology of additive printing of 3D objects is very exciting, the rise of Computer Numerical Control (CNC) manufacturing is also hypnotizing. If you're like me, the warmth and beauty of stone or wood is something that might never be replaced by a 3D printer working in man made materials.

A CNC machine is a device that can carve any of your ideas into reality out of any material. Using CAD/CNC systems there is essentially no limit to the amount of detail that can be included in any construction project today. Prices of materials will drop considerably as we move into this space oriented century. We can make or print anything to house our families. Every home will be dressed like an estate!

Modular housing ideas have been, and continue to be, introduced to the American market with regularity. Reviewing history about these "solutions" in modular housing you find very little that speaks to a broad and lasting success in the market. The most successful example has to be the trailer home. Mobile. In other words, most people end up talking about the time they lived in the trailer park and don't plan to return if given the choice.

Walking around in my mobile home I always knew it was a place I wanted to be "from". There is almost no trim you can put on a double-wide where it doesn't always look like a double-wide. Americans are individualistic and our housing has always been an extension of our imaginations. There is no way that a modular design that appeals to the masses in China or France is ever going to be successful in our market.

The truth is that all houses are based on groupings of various cubic spaces. The thing that sets them apart is the way these spaces are connected and dressed inside and out. By using the imagination of the human mind and the power of the computer to drive the CNC routers or 3D printers a unique vision will surface. Each individual wall, each panel, any item, will be a projection from the owner's imagination. Marketing MBAs will no longer decide what you're allowed to desire.

CNC routers can cut the negative space defining the eventual exterior appearance of walls into medium density foam. This foam can then be the mold for casting intricate designs. Faux brick, stone, even hand tooled wood, or any imagined form! Concrete will be stained and poured onto the foam bases creating unique and one-off designs. Once dry, the mold is power washed off and the final finishing stains applied. The engineered and designed foam forms could be delivered to the customer site where they are poured and stained. Because we are using computerized design tools we are free to design what we want as long as we stay within the parameters of the material we are casting.

Unlike traditional tilt wall construction, these custom computer etched forms can have irregular edges where they fit together like zippers. Removing the vertical seam takes out the major source of monotony in current tilt wall designs. While geometric shapes with hard edges are most commonly envisioned when working with concrete, the use of the CNC routing of foam molds allows a new kind of creativity to bloom.

Go to Youtube.com and search "CNC sculpture" for an astounding collection of videos demonstrating the current wave of CNC machines. If watching a 5-axis CNC machine quickly sculpt a human head from aluminum doesn't free up your imaginative juices then you really aren't paying attention. There is no building face, no design feature, that can't be quickly and accurately modeled.

Can you imagine a computer controlled machine building a form for a rock wall where every stone face is unique, just like real masonry? Where the color of every stone is uniquely stained? Where every simulated beam in the wall has unique cracks and grains like real lumber? Why not?

We'll be able to choose from history of world architecture for our ideas. The most detailed work is possible without being negated by labor costs. All of us can take advantage of the history of ideas and live with and enjoy that very same beauty and style from anywhere in history. Your heart's desire!

Taking advantage of these technologies gives us the ability to rapidly produce totally unique building structures. As the library of files defining these exteriors, and interiors grows, we will begin to see the special character of EVERY homeowner start to live in our neighborhoods. Because the pre-fabricated pieces can be constructed in an assembly-line facility the speed and efficiency of the process will geometrically improve. In fact, the notion of the CNC design factory coming to the building site is already gestating.

Your vision is possible and getting closer by the day! The subdivision that consists of ten or fifteen models is a thing of the past.

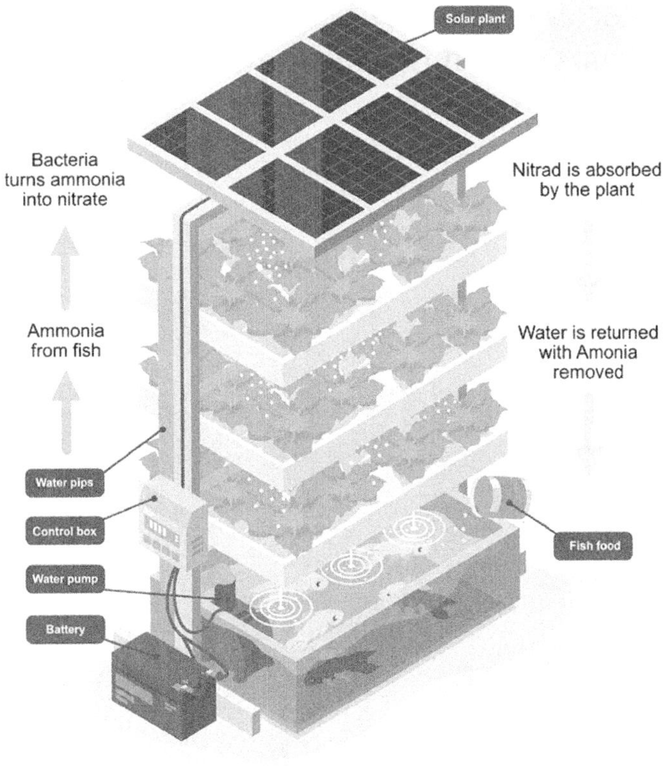
Solar plant
Bacteria turns ammonia into nitrate
Nitrad is absorbed by the plant
Ammonia from fish
Water is returned with Amonia removed
Water pips
Control box
Fish food
Water pump
Battery

Aquaponics

Aquaponics is the conjunction of aquaculture (the cultivation of aquatic animals in a controlled environment) with hydroponics (cultivation of plants in liquid nutrient). The interesting thing about this mix is that the waste products the fish produce is **exactly** the type of nutrients that the plants adore. The nitrogenous wastes from the fish eventually poison their environment. Plants thrive on these very wastes. The fish feed the plants and the plants groom the water for the fish. A perfect match.

I'm sure you're asking yourself how this fits in the new suburban environment. A quick search of the internet for "Urban Farming " will give you more to review than you can imagine. However, I come from a long line of meat eaters. Fish, and especially prawns, are something that fills that void for me. I simply won't consider a plan that requires me to live on rabbit food.

Building the fish tanks underneath the building protects them. In this way, the footprint of the system retains the dimensions of a small city lot. In addition, by building the fish ponds under the building you protect them from some of the environmental contamination and temperature extremes.

Large and small birds of every feather enjoy a bath in open ponds. If you've ever had a small goldfish pond and seen a great blue heron dining on your pets, or had a flock of geese decide to fertilize your patio, you know how quickly a large bird can change your environment. Keeping birds out of your ponds means fewer problems. The simplest solution is to prevent contamination in the first place and deny environmental factors from having the ability to contaminate your water source.

As a simple, incomplete, test of the 21st Century Homestead idea let's run a few numbers to see if we really can feed a family. The initial review will consider three 10'x20' tanks placed under a 30'x16' wing of the building. We're going to assume these tanks are 5' deep. By making the tanks just a few feet wider than the basic structure on the side, not much wider than a typical sidewalk, you have an access point to the ponds for feeding and equipment access. Having three systems isolated from each other you enable the urban farmer to raise several species, at different growth stages, and prevent some of the disease issues inherent in a large monoculture. Variety is also important!

Now the fun part where determining what kind of production we might expect from our suburban aquaponics systems. Once again, there is more information on "fish farming" via Google than you can process this week.

Aquaculture is all about weight of fish per volume of water and feeding density. Since there are 7.48 gallons/cubic foot and our tanks are 1000 cubic feet, that means each tank is almost 7,500 gallons. At the most dense stocking and management levels of the commonly used species in aquaculture, it is reported that you might be able to achieve a production level of 0.8 pounds/fish/gallon per year. This, of course, is assuming you are paying attention to the needs of the system.

When you multiply that 0.8 pounds by 7500 you start to see the beauty of this system. Each of our three tanks has the potential to produce 6000 pounds (3 tons) of fish per year. In other words, perhaps eighteen thousand pounds of fish per year, or on average 50 pounds per day.

One of my favorite species is fresh water prawn. A properly managed system might produce 0.4 pounds/gallon. This could conceivably mean that a single tank could generate around 8 pounds of prawns per day. With these numbers you can easily imagine that you could feed your yourself and it might even become a small source of income for your suburban family. In my particular case, I know it is impossible for me to tire of giant prawns. In any case, I'm willing to try and be bored by bacon-wrapped, jalapeno-stuffed, grilled prawns.

It is not a trivial task. It requires your complete attention. Like any agriculture, you can be surprised by circumstances. If you don't have the time to make sure the lawn stays mowed this is probably not a future for you. With this kind of production you can afford serious festivities. Grilling prawns with your neighbors is good fellowship!

Teach Fisherman

If you're the kind of person that wants to be given a fish to make it through another day you're totally wasting your time here.

The "21st Century Homestead" is all about allowing citizens to be capable, free, and independent. People really can be free to chase their dreams because they are unburdened by the worries about feeding their family. It's a core issue. It is not something that any real American wants to leave up to the whims of the government.

If you've been watching Washington lately you know there are politicians and sheeple that currently, today, see no problem spending large fractions of trillions...TRILLIONS of dollars to give people a fish. As Jefferson said so clearly, nothing could be more representative of a blot and open sore on the Republic.

Every fisherman knows, as does every farmer, that filling the larder requires hard work and patience. Another thing you need is determination. Attention to detail has a larger vote than pure luck. If you can't keep your grass mowed or your laundry done then the 21st Century Homestead will **not** be a boost to your success. If, on the other hand, you are able to manage standardized routines then I'd like to at least introduce you to some ideas and processes that can change lives starting today.

Hydroponics

This is the other half of the aquaponics equation and a most necessary part. The water treatment requirements of an aquaculture system are almost entirely subsumed by the plants in a hydroponics system. Intensive hydroponic systems under glass already provide hundreds of millions of pounds of fresh produce to the world's markets. Much of the bagged salad vegetables you find at the grocery are currently produced in this manner. The point is that it is a necessary part of the homestead ideal and something that is entirely scalable.

While almost any garden vegetable can be grown in the aquaponics regime, the most commonly used are leafy vegetables. Rapid growth implies a more rapid removal of the fish waste from the water but it also means more routine harvest of sustenance available to the urban farmer. Being from the south, when I think of fish, I often associate it with peppers and okra in a gumbo. A recent experiment showed that okra grown in an aquaponics system produced fourteen (14!) times as much okra as that grown in a field of comparative size. If you've ever been responsible for picking the okra in any average size planting you know that is a humongous amount. Okra is prolific. That is the reason that large areas of Africa depend on it as the staple crop.

In the simplest terms, the aquaponics plants are grown in gravel with the fish tank water pumped over it. Because of this, it lends itself to entirely new perspectives in gardening. As long as there is ample sun exposure, the planting could become part of a pitched roof or even columns scattered around the yard. Sunshine and water bathing the roots from the fish ponds is all that is required. It sounds pretty simple except for the fact that it absolutely must be organic. Using pesticides on the garden directly impacts the fish. This is a feature!

Okra. Peppers. Squash. Lettuce. Cabbage. Tomatoes. Beans. Melons? Coffee? Tea? Your imagination and future plant breeding advances will lead to the Jeffersonian ideal of the honorable farmer...even in suburbia!

The "Socrates Project" was started by President Reagan. He realized we were being surpassed by other countries because we had switched from technology-based planning, common during WWII, to economics-based planning. We need to get back to technology-based planning.

Because the 21st Century Homestead idea is based on a modular building system it supplies an open framework. Imagine, please, that every year a graduating class of architects is encouraged through cash prizes for the most interesting and productive solutions based on this framework and utilizing 3D printers and CNC machining. You could donate to this today. In very little time at all we'll be sitting around steaming bowls of home-grown gumbo picking out our next project to download, print out the 3D object, and try it in our own 21st Century Homestead!

It's not quite the Star Trek replicator but it's closer.

Social Security

Now that's a couple of words that have a different meaning to so many people. If you're elderly they may represent a lifeline. If you're my age it represents a tax on my earnings that fund a government program in which I have some expectation of participation after a lifetime of paying. If you are my kids they have doubts if it means anything more to them than a tax on their earnings for a program that cannot possibly exist in its current form when they get to retirement age. I think most people can believe that the very meaning of "retirement age" will soon be changing.

You only have to watch the politicians playing with the social security tax in these times of economic strife to understand that it is a program which, once again, puts you at the mercy of others. It doesn't take a genius to realize that a government running trillion dollar deficits might be placed in a position of cutting back on your monthly check in order to save itself. You don't have to be too paranoid to realize that a payment generated by computers could be delayed by some kind of natural or man-caused disaster!

It is not security if your life depends on the whims of politicians in Washington, D.C. If you are one of those people that insists on calling Social Security an entitlement, an annuity that you've been forced by government to purchase, you cannot possibly continue to believe that your future is in good hands. Every solution to the Social Security funding crisis involves reducing benefits, extending retirement age, or a combination of those two. If your "entitlement" is subject to the whims of politicians then you really don't have an entitlement. You have a promise from a politician who thinks they know better than you how to manage your money. If you insist in believing this, ask the politicians for your account balance and not just what they say they'll pay you when you retire. The fact that your family has no social security account when you die proves that you are a wage slave for the politicians.

When you start poking around on the internet investigating the Social Security system as defined in the United States you notice the Ponzi scheme nature. It supplies the backbone for old age and disability assistance that has become something of a fossilized bone in the throat of government. In dollars paid it is supposedly the largest government program in the world and a huge part of the United States federal budget. But, like any Ponzi scheme, it is destined for eventual failure. One writer termed it "irreversibly insolvent". We are beginning to see the end. New workers supporting the system cannot cover the expenditures for new retirees. Failure is the near future. If you believe there is a political solution to the problem of unsustainability and envision ordering the government to print you more money as a possible solution, you're waking to reality is going to be abrupt.

The ideas in "21st Century Homestead" represent a truly independent, sustainable, safety net for gaining control of our lives. There is nothing more important than having the tools to feed and house your family.

Nothing makes you feel more secure. Jefferson would be proud to see our society based on a technology that returns people to their own soil and industry.

People can argue about what route one might take to build a safety net for true social security or social justice. Some say it is better to let the government invest your retirement savings and convince you to believe they really are investing your money once it lands in the hands of politicians. Others can point to the fact that the money is not really being invested as proof that you'd be better off doing actual investments yourself. You will lose freedom of choice in one plan and be free to be a loser in the other.

Another option is to spend a lifetime building your home and becoming proficient at feeding yourself regardless of the political or economic storms. Government could help by letting the younger generations invest their money in their future security inside a 21st Century Homestead.

Is that you?

The world market for affordable housing is huge. Whenever you start on a project you really have hopes that it will catch the imagination of people and, in internet parlance, go viral. The idea of a 21st Century Homestead is something that could very well approach this goal. The technology exists and the spirits are willing.

This idea has more application in the world than just gaining market share in the United States. This country is the largest charity giver in the history of the world. Billions of dollars are doled out in order to feed the children around the world. Give them another fish? You only have to look at some of the charities to feel like there might be another option. You don't have to dig into any charity too hard to find people making six or seven figures. How many could YOU feed for that kind of money?

If the charity focus of the philanthropists were changed to give the people a home that actually feeds them then the spread of true social security would become real. The best part is that instead of sending checks to some administrative organization that is forever asking for more and more money, pouring aid into a black hole of foreign corruption in many cases, this project could collect money for homes. 21st Century Homesteads. Fixed price, rapidly deliverable, and each as unique as the owners. **Homes can be seen, and numbered, from Google earth. Satellite tracking of the efficacy of your charitable donations!**

Unlike other charities, the continual advancement of the homestead framework and it's supporting technologies will accelerate each year. The speed and efficiency of the robotic builders will drive forward the idea of the individual being capable of supporting themselves inside their shelter. Even in America, recent studies seem to imply that up to 25% of families harbor fears that they might have some difficulty putting food on their table in the current month. This number is meaningless in the era of Chinese coronavirus. Farming your site advances the human condition!

And finally, once the processes are in place, we won't be needing any $350,000 per year administrator to make a difference in people's lives. Like the

Amish today, and Americans generally in the past, we could have house-raising parties around the world. When you have a technology that allows you to build a serious structure in just days you are rapidly attacking one of the most serious issues for all societies. Quickly.

Nothing but good here.

Shoot a feather into the sky!

Typically you'd expect to be presented a request to donate to a website or pitch an investment opportunity. This troubled me because I had no idea where or if this was going anywhere. In trying to work this out, I had an adviser mention that many people believe the world is overcrowded anyway.

Is the world overcrowded? Are we running out of resources to feed the teaming masses? I don't think so. The perception and reality are two different things. When was the last time you heard the phrase "peak oil"? We are evolving. Daily. And we must keep the needs of people foremost in our equations.

A little test.

Texas comprises 268,581 square miles. Estimates for 2014 were that 7.161 billion people lived on the earth. I figure that comes out to 42 people per acre if EVERYbody moved to Texas. Looking back into Texas history, Mission San Jose near what is now San Antonio, was founded in 1720. It comprised roughly six acres and was home to 350. That is about 58 people per acre. Of course, being in Texas, the settlers at the mission had access to thousands of acres free range for livestock of all kinds. While that might be a little crowded in single family homes, it leaves a pretty empty world.

Another exercise in the carrying capacity of the planet is Manhattan. Wikipedia shows that it is 27.83 square miles. Tiling Texas with it would require 9,650 Manhattans. The city itself contains 2,683 acres of park land, so this exercise would distribute 25,890,000 acres of parks across the state. This is significantly more public use land than exists in the state at present. In fact, it's five times larger than New Jersey. Who'd be the Garden State then? The next step requires you to believe that Manhattan works as a city. If you do, then consider the population. The 1,628,706 people of Manhattan would become a horde of 15.7 BILLION people. Almost twice what they tell us is over population today. In fact, the entire planet could move to Texas tomorrow (please don't!), and they'd barely fill up half the state by Manhattan standards, and leave the rest of the planet totally empty.

The 21st Century Homestead lays out a plan to start meeting the goals of an expanding population. Being a Star Trek fan from my youth so many years ago, I can't imagine that somebody would consider comparing the depth of ranch assets in Texas to the boundless wealth from our coming space exploits. At the same time, I'm sure they will.

RI 3204
RI 3204

These were some of my thoughts as I wrestled with finishing this pamphlet.

Shoot a feather into the sky! What?

Now that's a crazy thought. What in the world could that possibly mean? Something like that crosses your mind when you're deep in thought about something else and you wonder where it came from. Who can say?

At times, something like this just flushes through your mind and then you forget. Another weird thought gone by. One of many. Boojillion.

But then a thought like that leads you down a weird path. Perhaps even an enchanting path. I'm hoping I can share the path that generated this thought, among others, and lit up an interesting morning for me.

The story starts with taking my notebook out to the hunting lease to try and decide how to close out this little pamphlet. Surely the clean, cool air of the Texas hill country morning would help me figure out where to go with it.

So there I am, laying in the floor of a plywood box blind as the sun starts to rise. No deer. In general, Texans in the south don't handle cold all that well but it really wasn't cold. Mid forties. It was the north wind cutting through my blind at thirty knots and chilling my irises and making my nose run like a faucet that was bothering me. I had decided to lay on the floor and give my face a little respite.

I noticed a turkey feather as I arranged myself on the carpet and dirt floor.

You always have to plan your hunts. Laying down on my back, I started deciding how I'd get up quietly to check for the arrival of any game, exactly which hand hold, the rest for my .308 Winchester handgun. This all leading to a trophy wall-hanger of a buck. Hunting dreams are always like that! Big dreams easily dashed by a misplaced handhold that creaks.

That handgun is a much better shot than I am an aimer. We shoot all the time at 400 yards which is a real confidence builder. Making a shot inside 150 yards is completely within reason. However, because it is a handgun and I'm not nearly as reliable as a rifle, I only take head and neck shots. Clean miss or a painless mortal wound.

But those are plans and dreams. Then I started thinking about the number of misses that I've counted up following this neck shot strategy. I knew my stats wouldn't be much worse shooting Wiffle balls, or feathers. You can instantly see in your mind how a high velocity Wiffle ball might wander across the pasture. Obviously the turkey feather had burrowed into my brain.

Feathers? What exactly would that look like? How would you load them? Would they get ten yards out of a .308 Winchester boost or just disintegrate completely?

You can chase that a little before you think there has to be another way to come at this feather shooting idea. This is the 21st Century! You need a completely different perspective!

If you really wanted to shoot a feather into the sky you'd, of course, do it from the Space Shuttle! Come at it from the other side of the sky!

Launching projectiles from the space shuttle just sounds too cool! Shooting feathers into Earth's upper atmosphere? A simple slingshot would work fine. Shooting feathers at earth and watching them slip into the atmosphere at orbital velocity! Is that not a great optic?

Would a feather burn up? Or float down?

Krogh's Principle states, roughly, that for any problem you're studying there is probably a system in nature that already represents the ideal of what you're trying to uncover in your research. That would surely imply that a feather would float down, right? It's designed by God to handle atmosphere and survive buffeting. To fly. Tens of thousands of feet per second orbital velocity has to be something else. Besides, at that speed, you really won't be able to watch a single feather.

Shooting a feather off the space shuttle with a slingshot would be much better. You could watch it travel with the naked eye for a long time. Now this is something that could interest me for the better part of a space day. If you're going to be plinking with feathers it would be much better to have a target. A piece of space junk floating alongside the shuttle would be the perfect target!

Shooting feathers off the space shuttle towards the earth at your space junk target is just plain too cool. If the feathers actually float down then the shuttle leaves a wake of fiery feathers as it flashes around the world. It would be quite a conversation piece to have a feather, probably a big old turkey feather, that was slingshot-fired off the shuttle. People might actually try to collect them. Sell them.

If there's going to be a market for unique reentry turkey feathers you'd have to have some way to authenticate them. Obviously. You can't just have anybody selling a turkey feather to bozos saying it came from the space shuttle. You know there would be space shuttle turkey feather counterfeiters. Experts in burning and disfiguring feathers!

To make it reasonable for people to trade space shuttle turkey feathers you'd have to have a 21st century way of authenticating them. Laser etched. That's the ticket! Serialized with the name of the astronaut who was firing the slingshot and including launch information. You can almost see advertisers lining up to have their laser-etched feathers float back to earth and offering free meals if they're presented at the business. Laser-etchd

QR codes with the complete history. Crazy enough?!

They probably do burn up. Even so, that's a cool looking scene. You're shooting feathers, accurately, from a slingshot at a piece of space junk and as the feathers start to enter the atmosphere they become little feather meteorites burning across the sky for just a second! That sounds like a fun shooting game!

What if they don't burn up? What if the millenniums of perfecting an air tool by the birds really does allow them to float down? Doesn't that mean something like a big soaring bird, perhaps an albatross, might make it, too?

We're talking about space here so you couldn't very easily launch something like an albatross. They have a twelve foot wingspan. That would be a sight! Can't you just see an astronaut configuring the NASA-designed albatross-sized slingshot from the space shuttle bay. 3-2-1 bird away!

On the other hand, this is space we're talking about here. This is the 21st century. If we're going to build a model albatross we don't have to make it to scale. We could make it much different since we're designing an albatross to test reentry. It could be as different and unique as our CNC space welders can make it. A model albatross made of metal.

Can you imagine that albatross with each wing being one kilometer across. Holy cow! That would be a test of space-based CNC welders building a 3-D metal albatross. Something that big you could probably carry some cargo back to earth, but what would that be? Plus, it's obvious that there aren't any kilometers-wide landing strips. You'd have to splash a cargo albatross down near the coast and that would anger the environmentalists.

What if the albatross itself is the cargo? Constructed of space age materials where each piece is a structural component but at the same time something else entirely. Thousands, millions, of parts, like standardized 2x4's, would be making up it's form. After an exciting reentry and landing the space albatross is just towed to shore and disassembled. Each landing would be untold wealth to the coastal community. The unimaginable wealth of the universe landing near **your** home!

That's the ticket for sure! And since we're building a model albatross of metal, we might as well build it from space-based resources. In reality, if you're going to be building something this big you're going to have to use space-based materials. I'm just sure we could make it from asteroid-mined titanium. Right? Shoot, if we're building giant titanium model albatross' from refined asteroid materials, we have to build more than one or it would hardly seem worth the effort.

The vision of row on row of titanium space albatross in various stages of completion near the asteroids started to fill my mind. All the robotic welders, sparkling quietly in space, working non-stop creating the next albatross to be launched back to earth. On arrival, disassembled to provide the materials for humanity. Perhaps they're actually using Krogh's Principle and actually making the beams and structural components like bird bones. Rigid. Strong. Mainly hollow. Something like that might lead to structural beams on earth that would be lighter yet stronger than anything we currently use.

The space albatross probably won't end up looking like a feathered beast. That just seems too weird. It will have several layers of skin. The inner skin to help strengthen the structure. Something like a titanium foam which while relatively thin and rigid, provides additional structural strength. When the albatross is disassembled, the metal foam is stripped off in sheets and used for roads. That's interesting.

Titanium foam roads? Sure, why not? They aren't impervious to water and you can't really see them leading to potholes. Maybe the tread design is on the road and our vehicle tires are like racing slicks putting more rubber to the road for control. I was seeing rows and rows of trucks lining up to put new super highways down as quickly as the sheets could be lifted from the trucks coming from the shiny titanium space albatross that was parked at the beach!

Is titanium shiny?

If those flocks of titanium space albatross' are moving across space towards earth it would just make a great view if they were shiny. A never ending stream of high quality materials slowly making their way back to earth. Once started, a steadily increasing rain of products.

Well, yeah! **If we're shooting flocks of space albatross to earth, why not crash a few down on Mars in a few places.** Pre-positioned, pre-refined, pre-delivered stores of materials for any colonists crazy enough to leave the sunny beaches of earth.

Now we're cooking with gas!

After all this, it is perfectly clear that in order to test reentry characteristics of a titanium albatross we HAVE to mine the asteroids. Which, of course, brings up the question of why we aren't. John Fitzgerald Kennedy inspired us to land on the moon in a decade. It wasn't because there was cheese. Elon Musk is taking us to Mars at this moment.

Over forty years since landing on the moon and we're still not mining the asteroids. We have all kinds of asteroid classifications, and know they have more materials that we can ever mine on earth and yet we aren't mining the asteroids. What is the real question here? Elon Musk is answering this question.

Every project that NASA has that isn't directly related to mining the asteroids is hereby canceled! Sorry. Playing at king is something that I learned as a child. If you're ever going to be able to test the reentry characteristics of a titanium model albatross shot off the space shuttle with a slingshot you're going to have to make the tough decisions. Done! Finally making a decision is so refreshing!

But this can't be the only problem. This is too obvious. Right? I mean there's really no way politics could be involved in something like deciding that the American people, the one's who landed on the moon, need to spend money helping other people appreciate their history instead of continuing to MAKE history. Is it really possible to get a hundred people together that would rather see a new world order rather than mine the limitless riches of space for humanity? It is absolutely outrageous to believe that we should let the government/corporate world chain us up before we reach the freedom and boundless resources available in space!

The thing I don't understand is what the new world order is going to look like.

Are we expecting Americans to reach for the middle with some dictatorship before we move on? This can't be happening. There's no way I want to watch some backward country pull us towards their history before we take this chance to explode into humanity's true future.

We can start landing the space albatross' and start feeding and housing the people of freedom. As President...

Whew...that was wild! I can't really say how long that wandered and ricocheted around my skull but I know when I got up to look there still weren't any deer. I think it must have been more than an hour. For the rest of the hunt I mainly just stared out into the pasture reviewing what I'd seen in my mind's eye until they came to pick me up.

When I got back to camp I told everyone that I was going to go back and write out the weird thoughts I'd had in the blind. They didn't think too much of it because I'm known for weird thoughts. I sat outside the little travel trailer writing in my notebook for a couple of hours trying not to forget anything that had crossed my mind. I'm sure I failed.

After scribbling this much down I ate some cold fried chicken and went in to take a nap. I went immediately to heavy sleep. After some time I thought I heard turkeys perting around the front door. I just knew it was my friends trying to entice me to get up, playing a joke on me, but as I listened it sounded better than I'd ever heard any of them run a turkey call.

I lifted the curtains expecting to see them laughing their asses off at me but there really was an entire flock of turkeys at the steps. Right at the door. I grabbed my .308 handgun and slowly squeaked out barefooted as they walked off and tried to bag a huge gobbler with a head shot at about sixty yards. Yet another close miss. They didn't fly off!!! I got some shoes on and walked around the trailer and bagged a young jake with a neck shot at about forty yards as he was scooting through the brush. He made great turkey and dumplings.

That was a day I'll always remember. I've had almost everybody tell me that this is ridiculous to include here but if you had a vision that was very clear, in color, and just plain fun...wouldn't you share it?

The NEXT 300!

The world is full of really big problems but I have the thought, probably because I'm a Texan and aware of "The Old Three Hundred", that a small number of dedicated and adventurous people can change the course of history. Today. The numbers of the United States' slide into the economic abyss are really beyond human comprehension. On 3/16/2011 the government reported that there were ONLY 479,000 housing starts per year and that was the lowest annualized number in almost 30 years. In 2020, in the midst of the Chinese coronavirus pandemic, the United States had an annualized rate of housing starts of over 1.4 million units.

Targeting one percent of the housing market doesn't seem so unreasonable when you can dry-in a custom home in days. Think what that Housing Starts report would look like if it said "starts" and the next column was "completed in 5 days"? I can believe it. Can you? Imagine it as reality! Free your mind from "a mile a minute". Don't be robophobic!

If you enjoyed this little diversion, please, share it with two of your friends. If there is any potential for spanning the needs of man and our place in the universe then you can help do it a couple at a time. Always remember that we were promised the riches of the universe! It is time to gather for harvesting instead of arguing.

Plato is purported to have taught that all the wars and battles in life were due to the desires of the human body. The freedom gained through the 21st Century Homestead producing urban farming abundance and internet education means that humanity will finally learn to be the perfect citizen that Jefferson so long ago envisaged.

He holds success in store for the upright, he is a shield to those whose walk is blameless.

Proverbs 2:7

Made in the USA
Monee, IL
05 November 2026

67352881R00049